The King of the
SPIDERS

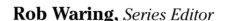

Rob Waring, *Series Editor*

T0175610

HEINLE
CENGAGE Learning

Australia • Brazil • Japan • Korea • Mexico • Singapore • Spain • United Kingdom • United States

Words to Know

This story is set in South America.
It takes place in the country of
French Guiana [giænə, -ɑnə].

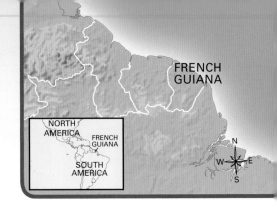

 Goliath Tarantulas. Read the paragraph. Then write the basic form of
each underlined word next to the correct definition.

    The Goliath tarantula is the largest spider in the world and can have a leg
span of over 30 centimeters when fully extended. It's native to the rain forest
regions of northern South America and usually lives in burrows in the ground.
Like some other tarantulas, the Goliath has sharp fangs and thousands of
barbed hairs on its body, abdomen, and legs. The Goliath also produces silk,
which it uses to make nests in trees or webs near the ground.

1. having sharp points that curve backwards: _____

2. the stomach area; the lower part of an insect's body: _____

3. a hole in the ground in which animals live: _____

4. a long, sharp, pointed tooth: _____

5. a delicate fiber or thread spun by spiders: _____

6. the measure of space from one point to another: _____

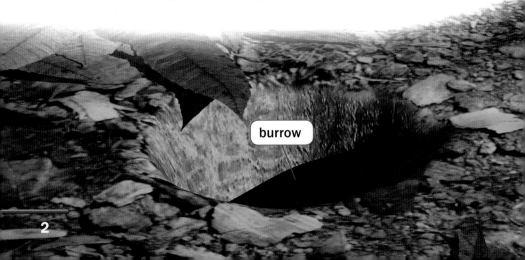

burrow

B **Tarantula Hunt.** Read the paragraph. Then match each word with the correct definition.

Tarantula expert Rick West has traveled into the dense rain forests of French Guiana to find Goliath tarantulas in their natural habitat. West's intention is to educate people about tarantulas since the spiders have often been given a bad rap. Although they carry venom in their fangs and have been known to bite humans when threatened, they are not usually deadly. However, their venom does prove to be a valuable weapon against their prey. The Goliath is a very dangerous predator for birds and other small animals.

1. dense _____

2. habitat _____

3. bad rap _____

4. venom _____

5. weapon _____

6. prey _____

7. predator _____

a. poison

b. a tool used to harm or kill

c. thick; crowded together

d. the area in which an animal lives

e. an animal that kills and eats other animals

f. animals killed for food by other animals

g. a negative general opinion about the quality of something which is untrue or without reason

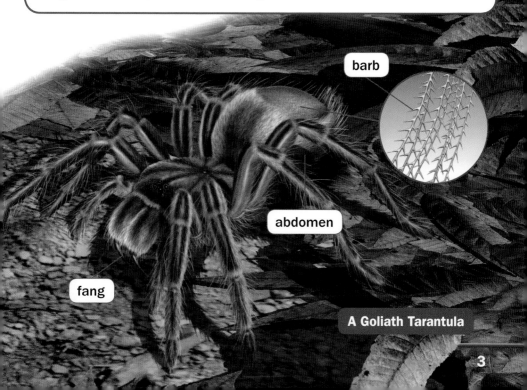

barb

abdomen

fang

A Goliath Tarantula

For some, the huge Goliath tarantula may be one of the most terrifying sights in the natural world. Witnessing one of these enormous hairy creatures **creeping**[1] along a tree branch can seem like a scene from a scary movie or a terrible dream. Its size alone may fill one with terror. This hairy spider, which is considered to be the biggest spider in the world, can reach the size of a dinner plate!

The Goliath is the largest of the tarantulas, and tarantula spiders in general are considered to be the biggest **species**[2] of spiders. There's an enormous number of different types of tarantulas out there, and they come in a wide range of shapes, colors, and sizes. The colors alone offer a huge variety—some are orange, others are blue, and still others are delicate shades of pink. In fact, there are more than 800 known species of tarantula in the world, and scientists are finding new ones all the time.

---

[1]**creep:** move slowly and quietly
[2]**species:** a specific group of living things that have similar characteristics

 CD 1, Track 01

Most people would prefer to stay as far away as possible from hairy monsters like the Goliath, but tarantula expert Rick West is not like most people. He's a man who is fascinated by tarantulas, and who regularly travels to the jungles of French Guiana to find the world's largest spiders in their natural habitat.

These amazing creatures don't seem to be easily frightened, not even by humans. Amazingly, they actually seem comfortable coming very close to people, sometimes even walking on them. West explains that tarantulas are often able to do this because they don't seem to be aware that they're walking on anything different. To them, a person could be anything from the natural world, even a tree. "They'll just walk on you," West explains as a large tarantula works its way across his shoulder and up towards his neck, "because they don't know me from a piece of wood."

Scan for Information

Scan pages 9 and 10 to find the information.

1. How many years has the tarantula been evolving?

2. How long can a Goliath tarantula's fangs grow?

3. Where is Rick West from?

The tarantula is truly the 'King of the Spiders.' It's a skilled hunter, which can jump incredibly quickly to catch smaller insects. Sometimes, it will even take on larger prey, including snakes. When hunting, tarantulas often wrap their hairy legs around their prey to hold it before **injecting**[3] it with venom from their long fangs. These large spiders are also tough survivors and are able to do well in almost every climate and landscape. Tarantulas have existed for an extended period of time, so they have had time to evolve and improve their skills. They're ancient hunters, the product of 25 million years of evolution, which has made them master predators, perfectly adapted to their habitats.

Of this incredible species of spider, the biggest and 'baddest' of them all is the Goliath tarantula. The Goliath can grow to over 30 centimeters* wide, with fangs up to 5 centimeters long. This huge spider makes its home in the remote rain forests of French Guiana, and that's where Rick West must go to find it.

---

[3]**inject:** force a substance (usually a liquid) into a person or thing
*See page 32 for a metric conversion chart.

The Goliath can grow to over 30 centimeters wide, with fangs up to 5 centimeters.

West is one of the world's leading tarantula experts, and his interest in the giant spiders often takes him from his home in British Columbia, Canada, to some of the most far-off and inaccessible places on the planet. It isn't an easy journey for him to see his favorite creatures deep in the French Guiana rain forest. First, he must travel via boat on one of the regions small channels deep into the jungle. Then, he must leave the boat behind and walk for miles even deeper into the forest, using a knife to cut his way through the thick ground cover.

Over the years, Rick has become a great supporter of tarantulas, and has devoted much of his time to trying to explain them to people all over the world. He feels that they really need his support as people are often biased against them despite having no reason to be afraid. One of the main reasons for this bias is the fact that the giant spiders usually receive a lot of poor and unfair coverage in movies and on television. "They have had a bad rap," he explains. "They've been **maligned**[4] in horror movies. These are the things that—as we've come up through our years watching television [and] science fiction movies—it's always the thing that creeps out of the shadow and **goes for the jugular**.[5] It **creeps people out**[6] and gives them the wrong impression."

---

[4]**malign:** speak or write bad things about someone or something
[5]**go for the jugular:** *(slang)* try to destroy someone or something
[6]**creep (someone) out:** *(slang)* cause fear and disgust

While tarantulas may not actually be the 'bad guys,' like some people think they are, they certainly live in some scary places. In his search for these huge spiders, Rick often goes into locations where most of us wouldn't dare to step, including dark, isolated caves in the middle of the jungle. "There's a cave here in French Guiana," West explains: "It is **primordial**.[7] As you enter down into this dark cavern and look backwards into the light, the vines are hanging down at the entrance."

As he talks about the cave, Rick also notes that it is not always an enjoyable place to visit. "You know there [are] snakes in there," he reports. "You can hear the bats starting to **swirl around**[8] as you enter the cave. You just don't know where they are, so that creeps you out." West then explains that, for him, exploring these caves and discovering the unknown secrets that lie within them is something he must do. The prospect of what he might find is irresistible. "It's one of those **compelling**[9] things. I have to go in to see what's in there."

---

[7]**primordial:** characteristic of the earliest beginnings
[8]**swirl around:** move in a circular or turning motion
[9]**compel:** force someone to do something

A Bat

Since Rick is in Guiana, he pays a visit to the primordial cave. He slowly cuts his way through the forest and finally reaches the cave entrance. As he nears the huge dark cavern, it seems as if he's walking into a black hole, but this doesn't bother West. He bravely continues to pick his way over the rocks and tree roots to get farther into the cave, using his flashlight to find his way. Once he gets deep inside, he begins overturning rocks and looking around. At last he sees what he has come for: a tarantula burrow. There could be a good-sized spider in there, so he cautiously approaches it and holds his flashlight towards the burrow in order to examine it closely.

After he's taken a look, West notes that there is indeed a tarantula in the burrow. "You can see a tarantula in it," he explains. "I [won't] know what kind it is until I get it out, so ..." With this comment, West bends towards the burrow and gets closer to the entrance so he can better see the tarantula. Then suddenly he says, "Wait a minute. It's come out the back. There's a hole in the back here." West quickly moves around to the other side of the burrow. "It's just on the back wall," he reports. "Maybe I can **tease it out**."[10] He then takes a long stick and gently pushes the tarantula towards the entrance to the hole. As the spider begins to move and comes into view, West says excitedly, "There it is, there it is! But he then adds with concern, "I don't know if I can get it." The spider is still quite deep within the burrow, one fast move and it could disappear once more.

---

[10]**tease (something) out:** carefully get something out of a tight place

Finally, West manages to get the tarantula out of its home and catches it in a plastic container as it runs along his leg. "That's a big spider," he says once he's placed it safely in the container. At that point, he takes a moment to clarify exactly what type of spider it is. "This is *Ephebopus rufescens*,"[11] he reports. "It's a tarantula that lives both in burrows in the ground as well as on cliff faces and in trees. And it's a new species." West actually helped to identify this particular type of tarantula on an earlier trip to French Guiana, but it's not what he is looking for on this trip. Now he is looking for Goliath tarantulas, so he opens the container and releases the spider back into its burrow. As he does so, the *Ephebopus rufescens* runs quickly back to its home and West heads back into the forest to carry on looking for a Goliath.

As West continues searching, one thing becomes very obvious; finding a Goliath tarantula in the dense jungle will be far from an easy task. He's well aware of the difficulties of getting through the jungle. As he pushes— and sometimes cuts—his way through the heavy plant cover, he describes the experience, "It's like barbed wire out there," he says. "I mean, there [are] some places where it's almost impossible to break through. You're always tripping over things. There's **razor grass.**[12] There [are] a lot of things in there that will cut and **scrape**[13] you."

---

[11]***Ephebopus rufescens:*** [ɛfeɪbəpəs rufɛsənz] the scientific name for this type of spider
[12]**razor grass:** a species of grass which has very sharp, knife-like edges
[13]**scrape:** cause a mark or injury by rubbing with a rough surface

Sequence the Events

What is the correct order of the events? Write numbers.

_____ West talks about the difficulties of the jungle.

_____ West releases the spider.

_____ West identifies the type of tarantula.

_____ West uses a stick to move the tarantula.

_____ West takes the tarantula out of its burrow.

A tarantula kicks off the urticating hairs from its abdomen.

West has more to worry about than just the challenges of finding his way through the thick jungle, though. If he finally locates a Goliath tarantula, he'll face a whole new set of dangers. Besides its fangs, the Goliath has some other very impressive weapons on its body. A tarantula's body is covered with tiny hairs, but it's the ones on the abdomen—called 'urticating hairs'—which can be dangerous.

Urticating hairs are one of the primary defense mechanisms used by some tarantulas. These barbed hairs cover the surface of the tarantula's abdomen. When threatened, many tarantula species will launch these barbed hairs into the air, directing them toward potential attackers. They do this by scratching the hairs off their abdomens with quick precise movements of their back legs. These hairs can sink deep into the skin, eyes, or nose of any animal that comes too close, causing the victim a great deal of pain and **irritation**.[14] It's surprising that these tiny hairs, not the fangs, are the first defensive weapon to watch for when approaching this spider.

---

[14]**irritate:** make something sore or painful

As night falls on the rain forest, West's search for the Goliath continues. As he slowly creeps along the darkened path, one must wonder if even an experienced spider hunter feels any fear. The jungle at night would be frightening for most people, but the darkness doesn't seem to scare West. For him, nighttime is a wonderful time to be in the jungle and he feels that it's 'electric,' or very exciting. Hearing noises but not being able to see where they came from—or even what made them—is actually thrilling for this expert. "The jungle at night, it's something where it's electric for me," he says. "You go into the jungle and you hear little noises ... **skitters**[15] in the leaf litter, and I'm telling you, it's **a rush**."[16]

After searching most of the night, West finally comes upon a tree that looks like it might be the home of a Goliath. He moves forward to closely inspect the tree, its leaves, and the plants around it. As he does, he notices a small circular hole near the base of the plant. It looks like there should be a Goliath there, but where is it?

---

[15]**skitter:** a quick movement
[16]**a rush:** *(slang)* a thrill

Suddenly, West hears a skittering noise on the jungle floor. He swings the light of his flashlight down towards the ground, and there it is, a huge Goliath spider. As he uses a large leaf to reach down and gently pick up the monstrous spider, he talks about his find, "This is a beautiful female. She is a monster. This is **Theraphosa blondi.** [17] It's the world's largest tarantula and the world's largest spider." As he holds it closer to his flashlight, the huge spider starts to walk across West's chest. "She's got to have a leg span of about ten inches across," he comments. When the spider begins to move even more quickly towards West's face and neck, he carefully puts his hand over her to block her progress saying, "[I] just have to slow her down here." When West moves his hand away, the spider calmly rests on his chest, while he has closer look.

A few moments later, the extremely large creature continues its exploration of West's upper body and slowly starts to creep over his shoulder. West remains perfectly calm. "Unless you really do something to injure it or to scare it," he explains, "they'll just walk on you. I'm not frightened at all by having a spider that size on my back." West allows the spider to climb over his back and down his arm, all the while examining the tarantula and observing its behavior. "She's got a lot of urticating hairs on the abdomen," Rick notes. "If she were to use her rear legs and scrape those hairs off her abdomen, they'd get into my skin and cause great irritation. So, I'm just going to [let] her go." And with that, West carefully sets the enormous Goliath back on the jungle floor and the tarantula moves quickly into the dark night.

---

[17] **Theraphosa blondi:** [θɛrəfousə blɒndi] the scientific name for this type of spider

Nighttime is the time when the Goliath is most active. They're 'ambush predators,' which means that they will make a trap for their prey, and then wait for a victim to fall into it. In the case of the Goliath, the trap is made of silk. Near the entrance to its burrow, the tarantula carefully lays down some silk. The silk is soft, but its softness is not to be trusted. The truth is that the silk of the Goliath tarantula is extremely dangerous—even deadly—for any small animal or insect that is unlucky enough to come across it. When the unsuspecting creature steps on the silk string, it moves slightly and acts as a trip wire, which lets the Goliath know something has come within range. Then the Goliath attacks!

Even though Goliath spiders have eight eyes, like most spiders, they have weak eyesight. This means that they often do not actually see their prey at all, but instead 'feel' that the animal or insect is there when slight movements travel across their sensitive hairs.

Back in the jungle, it's only a matter of time before some unfortunate creature approaches the Goliath's trap; this time it's a floor mouse. Little by little, the mouse moves closer to the tarantula's burrow, and then finally comes too close, touching the silk. It's like ringing a dinner bell for the hungry predator. The tarantula instantly jumps and wraps its legs around its victim, rapidly injecting it with its venom. The mouse has no chance against this master predator and dies very quickly.

While the Goliath may be able to kill a mouse without trouble, it's worth remembering that for most people, the tarantula's bite is no worse than a bee **sting**.[18] It may be painful, but not deadly. According to researchers, there has never been a single confirmed human death from a tarantula bite. Fact contradicts legend here, and despite its bad rap from television and science fiction movies, the tarantula is not a killer of humans.

As West watches the tarantula enjoy her meal, he prepares to wrap up for the night. He must go back to Canada as he has an important story to tell. He needs to spread the word that, although many human beings are scared of tarantulas, there's no reason for them to be frightened. Perhaps knowing the facts about these ancient predators can help turn human fear into a fascination with the King of the Spiders.

---

[18]**sting:** a bite resulting in pain and swelling

Summarize

Answer the questions. Write or give a short informational report about tarantulas. Use information from the questions.

1. What is the history of the tarantula?

2. How do most people feel about tarantulas, and why do they feel that way?

3. Why is the Goliath famous and how does it kill its prey?

After You Read

1. From the information on page 4, it can be inferred that tarantulas are known for being each of the following EXCEPT:
   A. hairy
   B. large
   C. new
   D. creepy

2. What is the writer's first impression of Rick West?
   A. His interests are not typical.
   B. He lacks intelligence.
   C. He has ambitious goals.
   D. He is distinguished.

3. The word 'remote' in paragraph 2 on page 9 is referring to:
   A. damp
   B. compact
   C. distant
   D. humid

4. Which of the following summarizes what Rick says on page 10?
   A. Humans usually treat tarantulas poorly.
   B. The Goliath tarantula is worse than fictional tarantulas.
   C. Some species of tarantulas occasionally kill humans.
   D. People are generally misinformed about tarantulas.

5. In paragraph 2 on page 13 'them' refers to:
   A. tarantulas
   B. caves
   C. vines
   D. bats

6. Which of the following is a suitable heading for paragraph 2 on page 14?
   A. Tarantula Teased Out of Burrow
   B. Rare Tarantula Escapes
   C. Goliath Found in Guiana
   D. Burrow Not Deep Enough

7. Rick needs a tool that can cut, _____ he can move through the thick jungle.
   A. which
   B. why
   C. that
   D. so

8. Where is the first weapon one has to watch for on a tarantula located?
   A. its head
   B. its back legs
   C. its abdomen
   D. its front legs

9. On page 20, the writer questions if spider hunters:
   A. ever feel frightened
   B. live too dangerously
   C. carry medicine for poison
   D. can see well in the dark

10. What will cause a tarantula to harm a person?
   A. if the person moves too quickly
   B. if the person hurts or frightens it
   C. if the person doesn't stay calm
   D. if the person tries to catch it

11. On page 24, the silk is described as 'not to be trusted' because it isn't:
   A. poison
   B. reliable
   C. silk
   D. harmless

12. On page 26, why does the writer point out that the Goliath can't hurt humans?
   A. to reinforce that tarantulas are not deadly
   B. to explain why the Goliath killed the mouse
   C. to prove that the Goliath is not the King of the Spiders
   D. to defend tarantulas against researchers

Tarantula Pets

In recent years, tarantulas have become increasingly popular as pets. For many people, caring for these creatures has become an enormously entertaining and educational hobby. There are over 800 different species to choose from and the majority of them are very easy to care for.

HANDLING A TARANTULA

When one is considering getting a tarantula, it is important to be aware that experts do not advocate handling them regularly. Not to protect the pet owner, but to prevent stress and injury to the spider. Tarantulas are not aggressive unless threatened by an abrupt motion, so gentle handling isn't usually a problem. If a person is bitten, the bite generally only causes redness and swelling similar to a bee sting. However, even a short fall can cause the delicate outer covering of the tarantula's body, its 'carapace,' to break, causing it to bleed to death.

CHOOSING A TARANTULA

Tarantulas can be purchased in pet stores, at pet shows, or even over the Internet. It is advisable to buy a spider only from a knowledgeable dealer who knows its exact scientific name and has already determined its sex. Knowing the species is important because each one requires slightly different care and feeding. Being aware of the sex is important because female tarantulas live much longer than males. A typical

Mexican Red Knee Tarantula

Tarantula Basics

1.	**Get a proper cage.**	A five-to-ten-gallon glass case with a tight-fitting cover is suggested. A screen cover is fine as long as it can be secured.
2.	**Cover the bottom with soil.**	Use one to three inches of clean soil. Some tarantulas, called 'terrestrial' tarantulas, like to dig deep into the soil and may require a thicker layer.
3.	**Create a hiding place for the spider.**	An empty flower pot turned on its side works well.
4.	**Add a water dish.**	It should be shallow to prevent drowning.
5.	**Attach a heating source.**	Try to keep the cage temperature between 24 and 27°C.
6.	**Get a food supply.**	Growing tarantulas should be fed live insects several times a week. A mature tarantula can last for one or two weeks without eating.

female can live for as long as 20 years, while most males die within a year or two. The most important aspect of purchasing a tarantula is making sure that it is healthy. A tarantula that is shrinking back in a corner with its legs pulled in under itself is probably dying.

CARING FOR A TARANTULA

For the most part, adult tarantulas require very little care. They should be fed once a week but can easily last for two weeks without any food. The cage requires only an annual cleaning. However, a tarantula regularly goes through a process called 'molting' in which it works its soft inner body out of its existing carapace and produces a new one. During the process, the spider's soft inner body is temporarily exposed and it can easily be injured or killed. It is essential not to disturb the spider during this process and, if there are signs that the tarantula is bleeding (its blood is pale blue), it may be necessary to coat the injured area with a hard substance to stop the loss of blood.

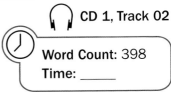

CD 1, Track 02

Word Count: 398
Time: _____

31

Vocabulary List

a rush (20)
abdomen (2, 3, 18 19, 23)
bad rap (3, 10, 26)
barbed (2, 3, 16, 19)
burrow (2, 14, 16, 17, 24)
compel (13)
creep (4, 10, 13, 20, 23)
creep (someone) out (10, 13)
dense (3, 16)
fang (2, 3, 7, 9, 19)
go for the jugular (10)
habitat (3, 6, 9)
irritation (19, 23)
malign (10)
predator (3, 9, 24, 26)
prey (3, 9, 24, 27)
primordial (13, 14)
razor grass (16)
scrape (16, 23)
silk (2, 24)
skitter (20, 23)
span (2, 23)
species (4, 9, 16, 19)
sting (27)
swirl around (13)
tease (something) out (14)
venom (3, 9, 24)
weapon (3, 19)

Metric Conversion Chart

Area
1 hectare = 2.471 acres

Length
1 centimeter = .394 inches
1 meter = 1.094 yards
1 kilometer = .621 miles

Temperature
0° Celsius = 32° Fahrenheit

Volume
1 liter = 1.057 quarts

Weight
1 gram = .035 ounces
1 kilogram = 2.2 pounds